THE SECRET WORLD OF

Flies

THE SECRET WORLD OF

Flies

John Woodward

Chicago, Illinois

© 2004 Raintree
Published by Raintree, a division of Reed Elsevier, Inc.
Chicago, Illinois
Customer Service 888-363-4266
Visit our website at www.raintreelibrary.com

Project Editors: Geoff Barker, Marta Segal Block, Sarah Jameson, Jennifer Mattson
Production Manager: Brian Suderski
Consultants: Michael Chinery, Marcus Stensmyr, and J. Christopher Brown
Illustrated by Stuart Lafford, Alan Male, and Dick Twinney
Designed by Ian Winton
Picture research by Vashti Gwynn
Planned and produced by Discovery Books

Library of Congress Cataloging-in-Publication Data
Woodward, John, 1954-
Flies / John Woodward.
v. cm. -- (The secret world of)
Includes bibliographical references (p.   ).
Contents: Six legs and two wings -- Flies of the world -- Getting around
-- Liquid diet -- Life cycles -- Flies and other animals --
Flies and people -- Flies and science.
ISBN 0-7398-7021-1 (lib. bdg : hardcover)
1. Flies--Juvenile literature. [1. Flies.] I. Title. II. Series.
QL533.2.W66 2003
595.77--dc21

2003002292

Printed and bound in the United States by Lake Book Manufacturing, Inc.
07 06 05 04 03
10 9 8 7 6 5 4 3 2 1

Acknowledgments
The publishers would like to thank the following for permission to reproduce photographs:
p.8 K.G. Vock/OKAPIA/Oxford Scientific Films; pp.9, 10, 16, 17, 28 Stephen Dalton/Natural History Photographic Agency; pp. 11, 15, 33 N.A. Callow/Natural History Photographic Agency; pp.12, 21, 22, 23, 27, 30, 31 Ken Preston-Mafham/ Premaphotos Wildlife; pp.13, 18, 20 Kim Taylor/Bruce Coleman Collection; p.14 Rodger Jackman/Oxford Scientific Films; pp.24, 35, 40 OSF/Oxford Scientific Films; pp.29, 32 Anthony Bannister/Natural History Photographic Agency; p.34 London Scientific Films/Oxford Scientific Films; p.36 David Tipling/Oxford Scientific Films; p.37 Alastair MacEwen/Oxford Scientific Films; p.38 Audio Visual, LSHTM, Wellcome Trust Photo Library; p.39 Daniel Heuclin/Natural History Photographic Agency; pp.41 Des & Jen Barlett/SAL/Oxford Scientific Films; p.42 Dr. Jeremy Burgess/Science Photo Library; p.43: Colin Milkins/ Oxford Scientific Films.

Other Acknowledgments
Cover photograph: Anthony Bannister/Gallo Images/CORBIS

Note to the Reader
Some words are shown in bold, **like this.** You can find out
what they mean by looking in the glossary.

Contents

CHAPTER 1
Six Legs and Two Wings

Flies make up a scientific order, or group, of insects called the *Diptera*, which means "two-winged" in Greek.

A blowfly like the greenbottle can detect the scent of its food—rotting meat—from a height of 115 ft (35 m) or more. If a blowfly were the same size as a human, it would be able to pick up scents from nearly 4 mi (6.3 km) away!

A fly can detect movement about ten times faster than a human. This explains why a fly is so hard to catch.

Flies belong to a large group of animals known as insects. All adult insects have six legs, and most flying insects have two pairs of wings, like a butterfly. But a true fly, like a housefly or a blowfly, has just one pair of wings. Many millions of years ago, the ancestors of flies had four wings, but over time the two rear wings gradually shrank into tiny stumps called **halteres.** On most true flies these halteres are very small, and often hidden, but they are actually very useful to the fly.

Many insects that have the word *fly* in their name, such as mayflies, dragonflies, and butterflies, are not really flies at all. They have too many wings to be true flies.

Just behind the fly's head is the **thorax.** Packed with muscles, the rigid thorax carries the fly's wings as well as its six jointed legs. The fly's tail section, or **abdomen,** contains the fly's internal organs (including most of the digestive system), and has up to eleven flexible segments.

▶ This is a fruit fly. It is a typical true fly and is related to houseflies and blowflies.

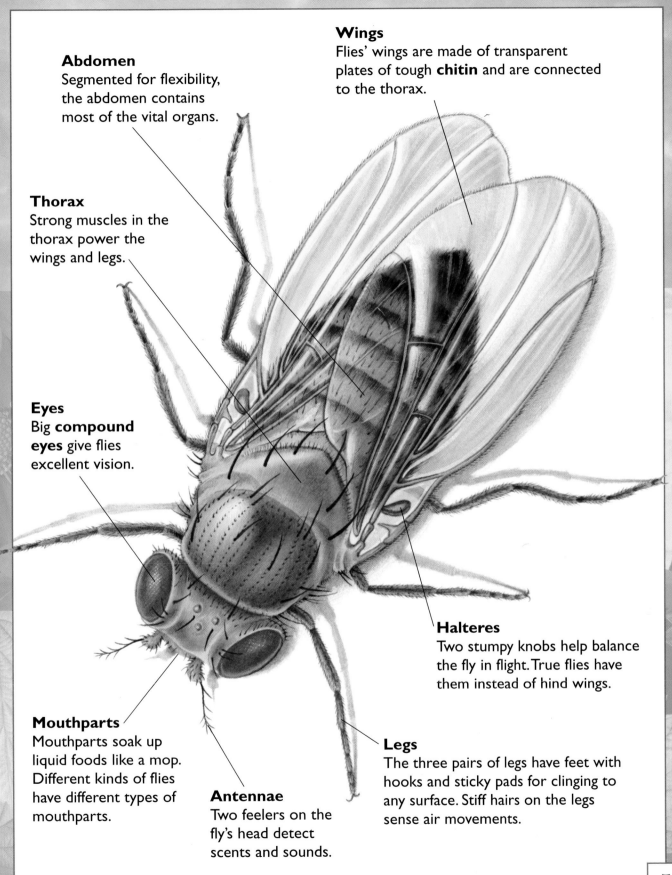

Abdomen
Segmented for flexibility, the abdomen contains most of the vital organs.

Wings
Flies' wings are made of transparent plates of tough **chitin** and are connected to the thorax.

Thorax
Strong muscles in the thorax power the wings and legs.

Eyes
Big **compound eyes** give flies excellent vision.

Halteres
Two stumpy knobs help balance the fly in flight. True flies have them instead of hind wings.

Mouthparts
Mouthparts soak up liquid foods like a mop. Different kinds of flies have different types of mouthparts.

Antennae
Two feelers on the fly's head detect scents and sounds.

Legs
The three pairs of legs have feet with hooks and sticky pads for clinging to any surface. Stiff hairs on the legs sense air movements.

ARMORED BODY

Even though a fly has one pair of wings rather than two, in other ways it is like most other adult insects. Its body is encased in a tough outer skeleton made of a substance called **chitin.** The skeleton is jointed like a suit of armor, so the fly can move. It has an especially flexible neck joint, allowing it to whip its head around with amazing speed.

AIR PUMP

Flies breathe air, like us, but they do not have lungs as we do. Instead, as in all insects, air is piped straight to where it is needed in the fly's body, through tiny tubes called **tracheae.** Air enters the tracheae through very small holes on the surface of the fly's body called **spiracles.**

In small flies, air flows easily in and out of their bodies. But some bigger flies can use their muscles to pump it in and out. If you watch a big, fat fly closely, you can see it breathing. However, neither of these methods would bring in enough oxygen to support a very big body. This explains why all flies, and most other insects, are so small.

The tough armor of this greenbottle gleams with beautiful metallic colors. The effect is caused by tiny ridges on the fly's body that break up and reflect the light.

GROWING PAINS

The fly's tough outer skeleton both supports and protects its body, but there is one problem with it: The skeleton cannot expand, so the adult fly is unable to grow. To get around this, the fly does all of its growing during an early phase of its life—not as an adult, but as a soft-skinned **larva,** or maggot.

A blowfly larva, for example, is just a fat, wormlike maggot with stretchy skin. As it grows, the larva molts (sheds its skin) twice, but it

This hover fly has landed on a sheet of glass, giving a clear view of its armored underside and its six jointed legs. You can also see the foot pads that allow it to cling to smooth surfaces.

is a quick and easy process. When it has finished growing, the larva turns into a **pupa.** Inside a protective case, the pupa's body transforms completely. Eventually an adult blowfly, complete with wings and six legs, emerges from the pupa shell. Once it reaches the adult stage, the fly stops growing and changing altogether.

The enormous, colorful eyes of a horsefly cover most of its head, allowing it to see in almost all directions. Each dot on the eye is a tiny lens.

lenses packed together like the cells of a honeycomb. Each lens can detect only the brightness and color of a tiny area, and the eyes cannot see in much detail. However, their vision is sensitive to movement.

A fly can also see **ultraviolet** light— which is invisible to humans—and is very attracted to it. Many people use electric fly traps that attract flies by producing ultraviolet light.

COMPOUND EYES

If you have ever tried to swat a fly, you will know how difficult it is to take one by surprise. Flies have extremely sharp senses, which they use to avoid danger, find food, and track down a mate.

Like nearly all adult insects, flies have very complex **compound eyes** made up of thousands of tiny

SCENT AND TASTE

A fly can sense chemicals in the environment around it using scent and taste. Its main scent organs are on its **antennae**—the short feelers on its head. They are tuned to pick up certain kinds of scents very well, such as the smell of its favorite food. When a fly lands on food, it can taste it using special sensors on its feet.

SOAKING UP THE SUN

The bodies of all animals work best if they are kept at the right temperature. If they are too cold, their muscles stop working. **Mammals** and birds avoid this because they are **endothermic,** or able to adjust their body temperature from within. But flies and other insects cannot do this.

They are **ectothermic** and must rely on the heat of the sun to keep warm. Some flies are able to live in the icy Arctic by sitting in white, cup-shaped flowers, which serve as natural sunrooms. Other flies shiver their wing muscles on chilly mornings to produce the heat they need to take flight.

Bristly Flies

A fly can sense movements in the air around it with the stiff bristles that extend from its armor. Each bristle is connected to special nerves, and if a bristle gets shaken by a breeze, the fly can tell where the breeze is coming from. It can sense sound vibrations traveling through the air in the same way. In this photo of a dung fly, you can see the stiff, black bristles sprouting from holes in the fly's armor.

I DIDN'T KNOW THAT

CHAPTER 2
Flies of the World

Flies live all over the world and in a variety of places. One kind of fly even has young that live in pools of crude oil.

Some flies are very tiny, with adults that are only 0.02 in. (0.5 mm) long. These include the scuttle flies that often live in compost heaps.

The biggest flies are crane flies. Their slender bodies can be up to 2–3 in. (5–7 cm) long and their wings can span nearly 4 in. (10 cm). Crane flies also make up the biggest fly family, with at least 15,000 different species worldwide.

There are about 100,000 known kinds—or **species**—of true flies. These include many insects we do not usually think of as flies. For example, the mosquito is a type of fly. Scientists believe there are many other species of fly that have not yet been discovered, especially in remote, tropical regions. In the cooler parts of the world, true flies makes up about a quarter of all insect species. True flies come in three main varieties: thread-horned flies, short-horned flies, and higher flies.

The bushy feelers on the head of this slender tropical mosquito show that it is a male. Male mosquitoes do not suck blood like female mosquitoes, who need the extra nourishment in blood to produce eggs.

The long-legged crane flies have longer halteres than most other flies. You can see them sticking out from the side of this crane fly's body, just behind its wings.

The thread-horned flies include bloodsucking pests like midges, mosquitoes, and blackflies. They are mostly delicate, slender creatures with long bodies, long legs, and long wings. The spidery crane flies, with their extralong **halteres,** are part of this group.

Although adult dung flies do not eat dung themselves, they do breed in wet cow dung. They also feed on other kinds of flies that visit the dung to feed.

houseflies and blowflies, as well as fruit flies, dung flies, warble flies, wasp flies, hover flies, and the wingless louse flies.

FLIES AT HOME

Flies live in a wide variety of places, or **habitats,** from hot, steamy rain forests and dry, parched deserts to the icy Arctic.

The boggy pools and streams of the far north attract thread-horned flies such as mosquitoes, midges, and blackflies, which lay their eggs in the water. When the **larvae** hatch they live

The short-horned flies are much less delicate than the thread-horned flies, and are often brightly colored. They include bee flies, insect-hunting robber flies, and pests like the biting horseflies. Most flies, however, belong to a group known as the higher flies. This group includes the familiar harmlessly in the water, but they soon turn into flying swarms of bloodsucking adult flies that plague the local animals and people. Vast areas of northern Canada, Alaska, Siberia, and Scandinavia are almost unlivable due to the number of flies that infest their swampy ground.

Nearly all adult flies live near the places where they lived as larvae. Yellow dung flies eat other insects, but they are always found near animal dung in fields because their larvae eat and live in it. Houseflies are most common in farmyards for the same reason, even though adult houseflies prefer to eat sweet, sugary foods rather than dung. Most other fly larvae live in the decaying remains of dead plants and animals, and so can be found almost everywhere.

Wasp Mimics

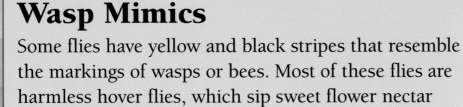

Some flies have yellow and black stripes that resemble the markings of wasps or bees. Most of these flies are harmless hover flies, which sip sweet flower nectar just like butterflies. But their stripes make these flies look as if they might sting. This is no accident. Many insect-eating birds avoid catching these flies just as they would steer clear of bees and wasps because of their stingers. This striped fly looks a lot like a wasp. Most birds—and many people—would avoid touching it. These kinds of flies are called **mimics,** because they copy, or mimic, the coloring of other insects.

15

CHAPTER 3
Getting Around

Because true flies have only two wings, you might think that they would be less efficient in the air than four-winged insects. Surprisingly, true flies are actually among the fastest and most agile of all flying insects. Many can fly in arrow-straight lines, switch directions almost instantly, hover in one spot, and even fly backward.

Flies are experts in the air, but they cannot fly upside down. When a housefly wants to land on the ceiling, it just reaches up with its front legs, gets a grip, and swings up to cling on with the rest of its sticky, hooked feet.

 The fastest flying insect on Earth is the deer botfly, which can reach speeds of 36 mph (58 km/h).

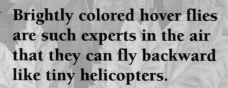 **Many flies have hinges at the base of their wings, allowing the wings to be folded back along the body. At the slightest sign of danger the wings swing forward, latch into place, and buzz into action, all within a fraction of a second.**

Brightly colored hover flies are such experts in the air that they can fly backward like tiny helicopters.

Some flies do not fly at all. Most louse flies do not have working wings and spend most of their lives clinging to the fur or feathers of animals.

WHIRRING WINGS

A fly's wings are simple plates of transparent **chitin.** They are connected to the top of the fly's thorax like the oars of a rowboat. When the top of the **thorax,** known as the **tergum,** is pulled down by the powerful flight muscles, the wings go up. Twitching another set of muscles makes the top of the thorax rise again, and the wings sweep down.

Since the tergum has to move only a tiny distance each time when the

Drone flies often hover in one spot for minutes at a time to defend their breeding areas. Hovering uses a lot of fuel, which explains why they have to keep visiting flowers to fill up on their sugary, energy-rich food.

fly twitches its muscles, it can go up and down many times each second, like a tiny, whirring motor. A housefly can beat its wings up to 200 times every second. But this is nothing compared to a mosquito, which can manage 600 wingbeats per second; or small midges, which achieve an amazing 1,000 beats per second!

Stabilized by its halteres, this little hover fly can hang in midair like a hummingbird as it investigates a willowherb flower.

halteres swing like tiny **pendulums.** Special nerves connect the halteres to the fly's brain. If the fly slips sideways in the air, its halteres swing a different way. The fly's brain instantly picks up the difference, and automatically corrects the error by changing the way the wings beat. Since this happens constantly, the fly is never thrown off course. The system also allows hover flies to hover in one spot without being blown off balance by a breeze. It works a lot like an airplane on autopilot.

AUTOPILOT

The wings of bees and wasps are similar to those of flies, but flies have something that allows them to outfly any wasp: the tiny, stumpy **halteres** that replace their hind wings.

Each haltere is like a short stick with a heavy knob on the end. As the fly moves through the air, the

MUSICAL WINGS

The familiar buzzing sound of a fly is the noise made by its beating wings. A small fly beating its wings 440 times per second produces the musical note A, the note that orchestras use to tune up. A bigger fly beats its wings more slowly and produces a lower note, while a smaller fly produces a higher note.

A small female mosquito buzzes with a whining F note, but when she is full of blood, the noise of her beating wings drops in pitch to D. Tiny, sensitive hairs on the **antennae** of a male mosquito are tuned to the whine of a female's wings. When its antennae start vibrating, the male knows there is a female nearby.

Hooks and Suckers

Like other insects, many flies can walk on almost any kind of surface. For example, a housefly like the one below can climb up a vertical sheet of glass with no trouble at all. It can even walk upside down on the ceiling. It can do this because each of its six feet has two sharp, hooked claws for gripping the slightest ridge, along with a pair of bristly pads like tiny suction cups. These pads are covered with a mass of tiny, hollow hairs that are kept moist—and very sticky—by a special fluid. The hairs on the feet are also sensitive to taste, so the fly can decide if its landing surface would make a good meal. Different types of flies have different types of feet, depending on the way they live and how they need to move around.

I DIDN'T KNOW THAT

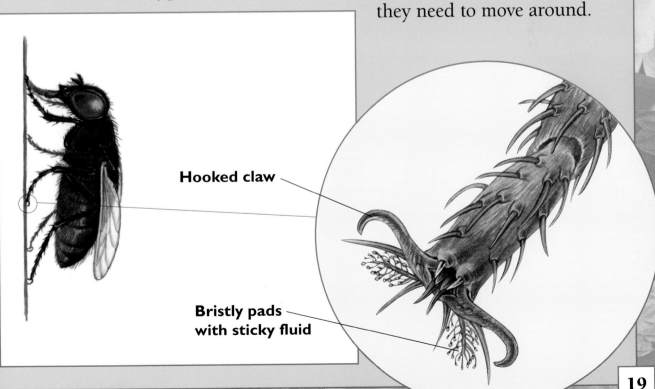

Hooked claw

Bristly pads with sticky fluid

CHAPTER 4
Liquid Diet

The mouthparts of many flies are made up of three to seven separate parts, which all fit together to form a tube.

A fly sucks up liquid food using a powerful pump, or in some cases two pumps, in its head.

Although bloodsucking flies are the ones that give humans the most trouble, the vast majority of flies around the world do not suck blood at all. They feed on dead plants and animals instead.

Only female mosquitoes suck blood. A single female may drink more than her own weight in blood at one time—but that single, giant meal may be all she needs for her whole life.

Like many insects, flies cannot chew their food because they do not have true jaws. They have to sip liquid foods, such as sugary flower nectar, or make solid foods into liquids that they can drink.

A fallen apple may look solid to you, but to a fly it is brimming with sweet juices. If it has started to decay, it is even juicier. The small flies known as fruit flies are particularly good at sniffing out the vinegar-like scent of rotting fruit or other decaying plant remains.

Blowflies have mouthparts similar to mops so they can easily soak up fluids. This bluebottle (a kind of blowfly) is feeding on the juices oozing from a piece of meat.

STRANGE TASTES

Many flies prefer foods that most people would consider very strange. These are the flies that feed on the liquids that seep from animal droppings or from dead animals' decaying remains. It may sound disgusting, but these liquids are full of nutrients that help keep bodies of flies healthy.

RICH DIET

Some flies are hunters. Robber flies and dung flies catch insects,

Gray mold has started to grow on this fallen papaya fruit and is helping to turn the flesh into liquid. The decaying fruit releases a fragrance that attracts swarms of fruit flies who come to feast on the sweet juices.

including other flies. Some flies even steal their **prey** from spiders' webs. Many more bite bigger animals to steal their blood. These flies include bloodsuckers like tsetse flies and louse flies, as well as female mosquitoes, horseflies, and blackflies.

TUBES AND SPONGES

Flies have a variety of different mouthparts for collecting their liquid foods. Hover flies have short tubes, which are ideal for sipping nectar from open flowers like daisies. Bee flies gather nectar from long, tubular flowers and have much longer feeding tubes, similar to those of butterflies. They often hover at flowers like tiny, furry hummingbirds, clinging

The bee fly's extra-long feeding tube is an ideal tool for reaching deep into an English primrose flower to sip the nectar.

to the petals with their front feet while they suck up the nectar.

A hunter such as a robber fly uses a stiffer, sharper tube to pierce its **prey.** It injects juices that turn its victim's flesh to soup, which it then can suck up into its stomach.

A mosquito does the same thing, but its needlelike feeding tube is so small that it can pierce skin almost painlessly, often allowing it to feed without being noticed. Horseflies slice into their victims with tiny knifelike mouthparts, then soak up the blood through tiny sponges attached to their mouths. Houseflies also have spongelike mouthparts to mop up their food. They often dissolve food in saliva first to turn it to liquid.

ENERGY FOR LIFE

Some adult flies do not need to eat at all because they eat so much when they are **larvae.** They store up enough energy to keep them going for a week or two. This is all the time they need to find mates and lay their eggs before they die.

Many flies are able to survive for longer by supplementing their stored reserves with nectar and fruit juices.

The sugar in these foods gives flies the high-energy fuel they need for flying.

Having captured a fly in midair, this robber fly is stabbing its prey with its sharp feeding tube and pumping it full of digestive fluid. Robber flies suck their victims dry, leaving just an empty shell behind.

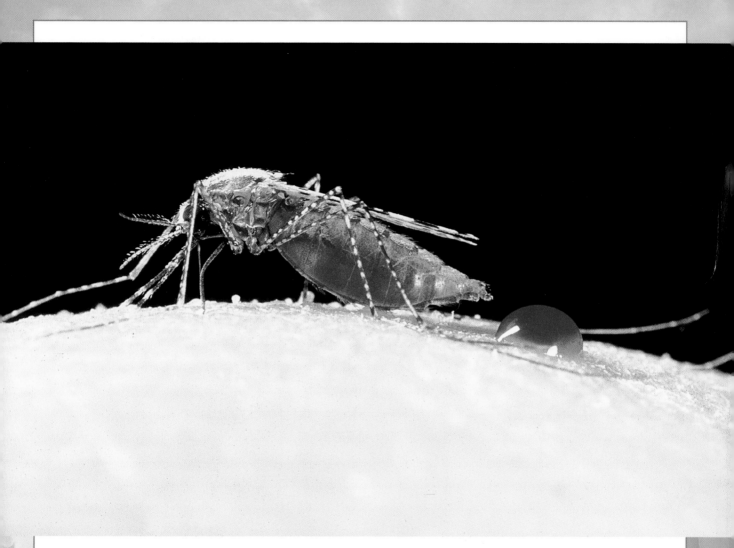

SUCKING BLOOD

With a few exceptions, blood feeding is usually done only by female flies. Blood is rich in the protein that female flies need to make their eggs. Since males do not make eggs, they generally feed on nectar and fruit juices instead.

IMPORTANT ROLE

Flowers often swarm with flies fueling up on nectar. As they drink, flies often get covered

The blood meal of this female mosquito glows red through her thin skin. As she is feeding through her needlelike mouthparts she gets rid of some of the excess water from her liquid lunch to save weight on take-off.

in flower **pollen.** They fly off to nearby flowers, carrying the pollen dust with them. If the dust brushes off onto another flower, it fertilizes the new flower and makes it produce seeds. So, like bees and butterflies, flies are important to the life cycles of plants.

Smelly Attraction

Many flowers attract insects such as bees and butterflies with wonderful fragrances, but a few specialize in attracting flies using the disgusting smell of rotten meat. One of the smelliest of all is the dead horse arum. With its hairy, brown, flesh-colored flowers, this arum is irresistible to blowflies looking for decaying flesh on which to deposit their eggs. The flies land on the flower and climb down into its warm, dark center to lay their eggs. Up to 40 buzzing flies may be trapped for several hours beneath spines inside the flower. If they have already picked up pollen from another arum they will **pollinate** the new flower. After one day, the spines inside the arum wither and the blowflies can crawl out again—but not before the flies have picked up a fresh dusting of yellow pollen to take to the next flower.

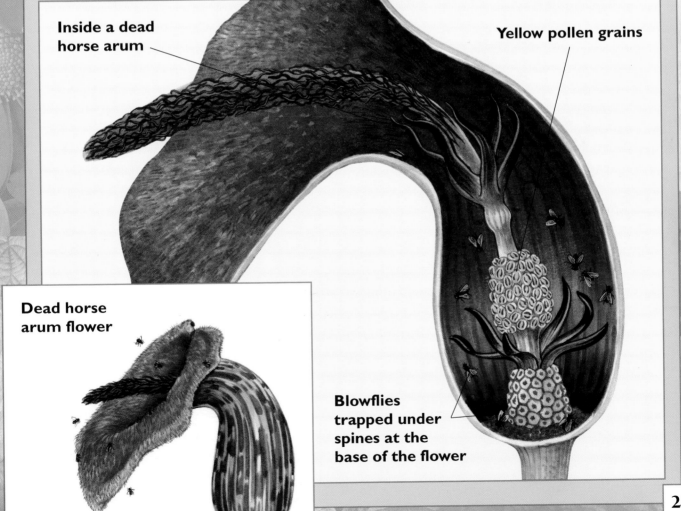

Inside a dead horse arum

Yellow pollen grains

Dead horse arum flower

Blowflies trapped under spines at the base of the flower

CHAPTER 5
Life Cycles

In warm weather, the entire life cycle of some flies, from egg to egg-laying adult, can take less than a week. Other flies may take a year to complete the process, particularly those that have to survive cold winters by lying inactive for months until springtime arrives.

When some female snipe flies are ready to lay their eggs, they gather in tight masses like swarms of bees. They produce their eggs and then die. When the eggs hatch, the larvae eat the dead bodies of their mothers.

When a blackfly is ready to emerge from its underwater pupa, it inflates the tough outer case with air until it bursts. The fly then shoots to the surface in a bubble of air, ready to take off and find a mate.

Young flies are very different from their parents. They are wingless, legless **larvae** with few senses. Many are just soft, wriggling maggots with a mouth at one end.

The mouth is the most important part of the larva. From the moment it hatches from the egg, a fly larva has just one aim in life: to eat. Its job is to gather food as fast as possible. As it eats, it grows. Its soft skin first stretches, then splits and peels away (molts) to reveal new, looser skin below. Eventually, when it is fully grown, it stops eating and becomes a **pupa.**

The pupa of a higher fly such as a housefly is surrounded by a case that looks like a hard-shelled seed. Inside, the creature's body changes into its adult shape, and hatches from the pupa as a fully formed adult fly. This process of change from egg to winged adult is called **metamorphosis** and it is common in many insects.

▶ A typical higher fly such as this blowfly has a four-stage life cycle. The adult fly lays eggs that hatch into soft, legless larvae. When each larva is fully grown it becomes a pupa. Inside the pupa, the larva changes into an adult fly with wings.

Eye-to-Eye

Rival male flies are often careful to size each other up before risking a fight over a female. Some tropical flies are perfectly equipped for this, with big eyes on the ends of long stalks protruding from the sides of their heads. They stand facing each other, like the pair you can see here. The male with the shorter eye stalks usually retreats and goes off to pick on another fly closer to his own size.

LIFE CYCLE OF A BLOWFLY

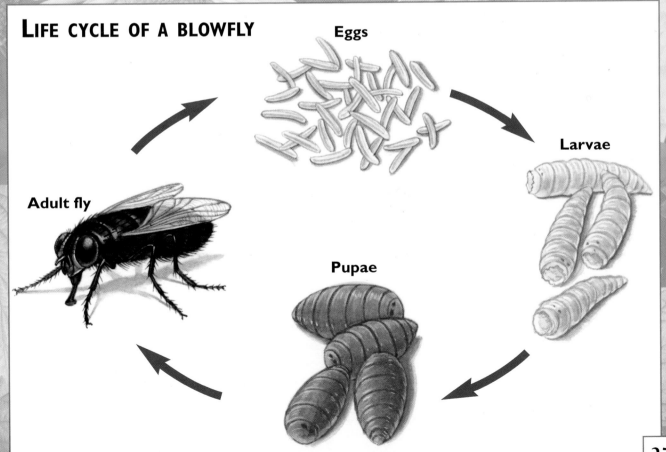

Eggs

Larvae

Pupae

Adult fly

LARVAL LIFESTYLES

Different types of fly have different types of **larva.** The larvae of houseflies, blowflies, and other higher flies are maggots, which live in wet places on land—or even inside other animals. Mosquito larvae have bristly bodies and well-developed heads with tiny mouthparts for chewing. They live in quiet pools, hanging under the surface from small floating tubes

The larvae of mosquitoes develop not on land, but in water. They can be found in huge numbers, hanging from the water surface in bog pools, swamps, and even the rainwater that collects in discarded car tires.

like snorkels. They feed on tiny edible particles in the water. The larvae of blackflies are similar, but they live in fast-flowing streams. They cling to rocks with special hooks and use two fans of sticky bristles to trap food particles.

VEGETARIANS AND HUNTERS

Some fly larvae are vegetarians. As you might guess from their name, carrot fly larvae munch their way through carrots, and the larvae of holly-leaf miners **burrow** through the outer layers of holly leaves, leaving pale blotches that show where they have been. Some larvae

are active hunters. For example, the larvae of some hover flies attack small insects like aphids. A single, hungry hover fly larva can eat up to 80 aphids per hour.

This tarantula spider was stung and paralyzed by a female wasp as food for her young. The body was then discovered by a female fly, who deposited her own eggs on it. The fly larvae will soon eat what is left of the spider.

WASTE DISPOSAL

Most fly larvae live in ways that probably seem disgusting to you. Many burrow into the bodies of other animals (such as insects, sheep, earthworms, and horses) and eat their flesh. These include pests like botflies and warble flies. Other larvae live in animal droppings and the dead remains of plants and animals, feasting on rotting tissues and **bacteria.** These larvae are **decomposers,** or organisms that help return nutrients from dead and decaying matter back to the soil. Without helpful decomposers like larvae, the world would be covered in decaying plants and animals.

MATES AND RIVALS

While a fly **larva's** only job is to eat, an adult's job is to mate and produce more larvae. Many adult flies barely eat at all, especially the males, but some use food to attract females. Male dance flies catch other insects, wrap them in silk, and offer them to females as "wedding gifts." Some fruit flies offer females fruit juice instead.

Other male flies dance in front of females and show off special patches of pale bristles on their legs, wing spots, or brightly colored body parts. One North American robber fly may dance for several hours to attract a female. Rival males also fight over females. Yellow dung flies often end up in struggling heaps as several males try to win over a single female.

EGGS AND YOUNG

Depending on the **species,** a female fly lays between 1 and 250 eggs in a single batch and up to 1,000 eggs during her entire lifetime.

Once a female mosquito has enjoyed her blood meal, she can use the rich food in the blood to make her eggs, and then lay them in a quiet pool.

Most female flies lay eggs that hatch into larvae. A mosquito lays her floating eggs on the surface of a pool, and sometimes sticks her eggs together to form tiny rafts. Blowflies lay clusters of eggs in animal corpses, and fruit flies inject their eggs into fruit or plants.

Some flies produce live young. Female flesh flies place live larvae on dead meat, often alongside blowfly eggs. These larvae may eat the blowfly maggots when they hatch. A female tsetse fly has just one offspring at a time, which she keeps in her body and feeds on "milk" made from the blood of

Dung fly larvae eat cattle dung, but the females must lay their eggs on the dung before it dries and gets too hard. As the females lay their eggs, the males cling to their backs to guard them from rival males.

the animals she bites. Instead of producing hundreds of eggs and abandoning them like most other flies, the tsetse fly carefully nurtures up to twelve young over the course of her entire lifetime. When each larva is fully grown, usually after ten to twelve days, the female tstese fly gives birth, dropping the larva onto the ground, where it immediately becomes a **pupa** and then turns into an adult fly.

CHAPTER 6
Flies and Other Animals

Mosquito attacks can be deadly. In 1963 about 200 adult cattle and 500 calves died from blood loss after being attacked by huge mosquito swarms in Louisiana.

The thick-headed fly sits on flowers and attacks bees. When it sees a bee, it flies up, grabs it, and lays an egg on it. When the fly maggot hatches from the egg, it buries itself into the bee's body and slowly eats it alive.

Some hover fly maggots live in ants' nests and eat the ant grubs. Amazingly, the ants ignore them, because the fly larvae produce a chemical that makes them smell just like ants!

The bee flies that feed from flowers lay their eggs near holes in the ground made by mining bees. The fly maggots crawl into the nests, find the young bees developing inside, and eat them.

Many flies are bad news for other animals. Caribou (a kind of deer) are tormented by huge swarms of mosquitoes in the Arctic, and large numbers of blackflies can suffocate cattle by blocking their nostrils and lungs. Yet the real villains are not the bloodsuckers, but **parasitic** flies. **Parasites** are animals that live inside other animals, getting food from them and not giving anything in return. Sometimes parasites even hurt or kill their **hosts** (the animals in which they live).

After feasting on a plump caterpillar, the larva of a tachinid fly emerges from its host. Only the empty skin of the caterpillar remains.

Link in the Chain

Although some animals may be bothered by flies, others rely on them for food. Orb weaver spiders are extremely fond of big, juicy flies, and swifts, swallows, and flycatchers eat them as well. The hover fly below will make a meaty meal for the spider that has first paralyzed it with poison and then wrapped it up in silk. Flies are also important in the diets of frogs, lizards, and bats. The fat **larvae** of many flies are a valuable food source for many animals and birds. Birds like crows and magpies will feast on the maggots they find on rotting animal corpses. Flies are, in fact, a vital link in the food chain that supports the huge variety of animal life on Earth.

Many adult flies lay their eggs on or inside other animals, so that the flies' larvae live as parasites. There are many advantages to this. The larvae are always warm and surrounded by their favorite food. They are in no danger of being eaten by other animals, and as long as they are not too greedy, their host animal will stay alive.

PROBLEM PARASITES

Some **parasitic** flies are deadly, especially when their victims are other insects. For example, the bristly tachinid flies sometimes lay their eggs in the mouths of caterpillars, grasshoppers, or beetles. The eggs get swallowed, and when they hatch, the **larvae** eat the unlucky **host** insects from the inside out, killing them in the process. Golden-furred cluster flies often kill earthworms this way.

Other flies target bigger animals such as rabbits, deer, sheep, cattle, and horses, living inside them without killing them. Some of the most damaging are warble flies and botflies. A warble fly lays its eggs

Cattle have learned to dread the hum of an approaching warble fly and they often run away in terror. This warble fly is a male, but an egg-laying female will not be too far away.

on the leg of a cow, and when the larvae hatch they **burrow** under the cow's skin. The larvae gradually make their way under the skin to the cow's back, where they feed on her flesh inside painful swellings called warbles. When they are fully grown the larvae chew their way out, drop to the ground, and turn into adult flies. The cycle then starts over.

Horse botflies make a different journey. This fly lays its eggs on a patch of hair that the horse regularly licks. The maggots hatch when they are licked, invade the horse's mouth, and then travel down into its stomach. They attach themselves to the stomach wall and feed there for up to ten months. They finally come out in the horse's dung. If there are enough of them, these maggots can make a horse very ill.

FLY-STRIKE

Flies are particularly dangerous for sheep. The sheep botfly places its live maggots directly into a sheep's nostrils and they crawl into its nasal passages and live there, making the sheep dizzy and unstable. The maggot stays in the nasal passages until it is fully grown, then it falls out (or is sneezed out). Sheep may also be attacked by blowflies that lay eggs in dirty wool. The maggots hatch and chew their way into the sheep's body, a gruesome condition known as fly-strike. When the maggots are fully grown and ready to **pupate,** they drop out of the sheep's flesh and burrow into the ground, leaving the sheep very ill. Sometimes sheep even die from fly-strike.

CREEPY CRAWLIES

Some of the most frightening-looking flies spend their lives clinging to birds and other animals, sucking their blood. Louse flies have strong legs and claws that cling to fur and feathers. Many have no wings. Both the male and female louse fly feed every five days or so, and although they take only a small amount of blood each time, it is enough to weaken a small bird.

The nest of these young swifts is infested with wingless louse flies. Although the flies may weaken the birds by sucking their blood, they will not usually kill them.

CHAPTER 7
Flies and People

The first attempt to build the Panama Canal was abandoned in 1889 because more than 20,000 workers died from malaria or yellow fever, diseases carried by mosquitoes.

Malaria was once widespread in the United States and Europe. Ancient Romans suffered from it, too, because their city was built on a mosquito-infested swamp.

Roughly half the world's human population either has malaria or is at risk of getting it. The disease is the biggest killer of children under five years old in tropical parts of the world.

The tsetse flies that carry sleeping sickness seem to be attracted to dark colors, perfume, and aftershave, so people who go on African safari tours are often advised to wear pale clothes and avoid using anything scented.

Sucking blood from humans is risky for flies because they might get swatted. Many mosquitoes reduce this risk by attacking at night. They track you down by your body heat and the air you breathe out. You do not notice that you have been bitten until the next morning, when you discover an itchy bump. Like all flies that suck blood, the mosquito pumps a fluid into the wound that keeps the blood from making a scab. The itching is caused by the body's reaction to this fluid.

Many diseases carried by mosquitoes are difficult to prevent or cure, so it is always best to avoid being bitten in the first place. This woman wears a protective net in a mosquito-infested area.

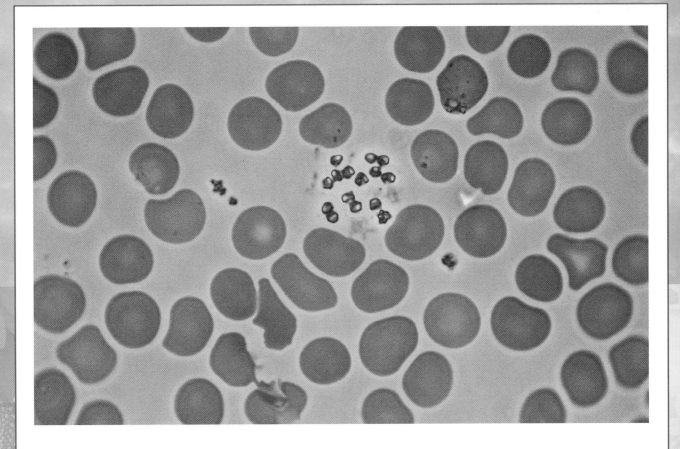

DEADLY DISEASES

The main danger to humans from bloodsucking flies is the infections they carry. Many of these flies—particularly the tropical ones—can harbor **microbes** that cause serious diseases such as malaria, yellow fever, dengue fever, Rift Valley fever, West Nile virus, hepatitis, elephantiasis, river blindness, and sleeping sickness.

Every year malaria kills two to three million people in the tropics, making it the deadliest of all diseases carried by flies. It is caused by a microbe that lives

These red blood cells (colored dark blue) have been photographed under a microscope. They have been invaded by malaria microbes, which have multiplied inside the cell in the center, causing it to explode. Further microbes released by the explosion will go on to attack other blood cells and cause the disease malaria.

inside mosquitoes and produces **spores** that get into their saliva. If these microbes are injected into a human by a bite, they multiply inside the red blood **cells** and destroy them. This causes severe fever, which goes away and returns again and again. If too many blood cells are destroyed by the microbes, the victim may die.

WEST NILE VIRUS

West Nile virus used to be unknown outside Africa. However, in 1999 the disease appeared in New York City. It may have been carried by a mosquito trapped in an airplane. Once it arrived, the virus infected birds, which then passed it on to local mosquitoes, which in turn passed the disease on to humans. Since then the virus has appeared in most of the United States and has infected nearly

The tsetse fly, like this male, is one of the few bloodsucking flies where both males and females feed on blood. Tsetse flies carry a disease called sleeping sickness.

4,000 people, almost 250 of whom have died. Now there are programs in place to keep mosquito populations under control, which has helped improve the situation. Although West Nile virus is still a serious disease, the chances of any one person getting sick from a mosquito bite are still very small.

OTHER PROBLEM FLIES

Mosquitoes are not the only disease-carrying flies. In tropical Africa, bloodsucking tsetse flies spread a deadly infection that affects both people and large grazing animals. It is called sleeping sickness because it invades the nervous systems of its victims and makes them drowsy and confused.

In Africa and Central America blackflies carry a disease called river blindness. It is caused by worms that infest the flies' saliva. When an infected fly bites a human, the worm moves into the wound and lives under the skin, where it causes intense itching. If these worms get into the eyes, blindness can result.

This boy is leading a man suffering from river blindness in Africa.

BACTERIA CARRIERS

Along with spreading diseases caused by **microbes** that actually live and breed inside their bodies, flies can also pick up diseases without being infected themselves. Houseflies and blowflies are well known for this because they feed on animal dung. They can pick up harmful **bacteria** on their feet and body bristles and then carry the bacteria into our houses. Flies can spread all kinds of killer diseases in this way, including typhoid, cholera, and dysentery. In very poor countries, this can be a deadly problem. Such places often lack systems for keeping communities clean, and flies can swarm around waste that flows in open sewers.

These houseflies are crawling over someone's sandwich. Flies can transmit disease-causing bacteria to humans by landing on our food.

UNWELCOME GUEST

A few flies will use humans as nurseries for their young if they get the chance. The most ingenious is the human botfly, which lives in the tropical forests of Central and South America. This fly glues its

Saved by the Fly

The tsetse flies that carry sleeping sickness have lived on the grasslands of Africa for thousands of years, and the native wild animals have become almost **immune** to the disease. When European farmers moved in and tried to raise cattle on this land, the disease proved so deadly to both the farmers and their cattle that they had to give up. This is one of the main reasons why huge areas of Africa's grasslands—with their native giraffe, springbok, zebras, and lions—still exist today much as they did long ago.

egg to the tail of a bloodsucking mosquito. When the mosquito attacks a human, the botfly maggot hatches and **burrows** into the wound. It anchors itself with rings of stiff bristles, and starts eating the human flesh. It grows bigger and fatter, producing a painful sore. When it reaches about 1 inch (2.5 centimeters) long, it drops out, burrows into the ground, and turns into an adult fly.

CHAPTER 8
Flies and Science

The fruit fly breeds at an amazing rate, producing a new generation every twelve days.

In 1933, Thomas Hunt Morgan was awarded the Nobel Prize for his scientific research work on genes and chromosomes, made possible by the fruit fly.

The university room where Morgan and his colleagues worked was called the *Drosophila* Laboratory, after the scientific name of the fruit fly, *Drosophila melanogaster*. But everyone else just called it the Fly Room.

▶ This scientist is using a microscope to examine fruit fly genes. The flies have been bred in the bottles you can see at the bottom of the picture.

Flies are not all bad news. They have also played a very important role in science over the last 100 years. In 1907 the American scientist Thomas Hunt Morgan started using flies to study **genes** at Columbia University.

The **cells** of all living things contain genes, which are a very complex kind of code. Genes are passed from parent to child, and determine what each individual will look like, as well as some of its behavior. Genes are so tiny that they cannot be seen by the human eye without the help of a microscope.

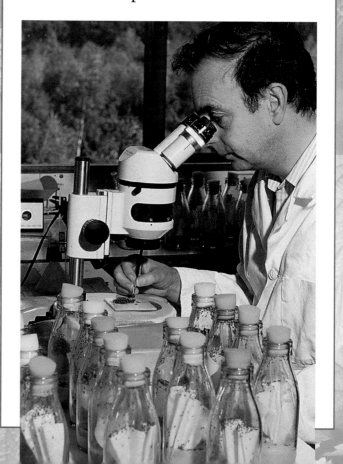

Most fruit flies have red eyes, but the genes that control eye color sometimes get damaged and produce flies with white eyes. Under the microscope, scientists have been able to find out which genes were damaged. This helped them discover the functions of different genes.

About 60 percent of the genes in fruit flies are similar to those found in humans.

Fruit flies were chosen for scientific research on genes because they breed particularly fast and many generations of them can be studied over a short period of time. Some of their **chromosomes** are also unusually large, allowing the genes to be clearly seen under a microscope. One of Morgan's students worked out the first "gene map," showing the function of each gene in a fruit fly.

By learning how genes work in fruit flies, scientists are now much closer to understanding how they work in humans. After research on human genes lasting ten years, the Human Genome Project produced its first human gene "map" in February 2000. Scientists hope that this map will help them combat many human diseases. Without the knowledge gained from research on the tiny fruit fly, this would not have been possible.

Larval Therapy

The thought of a mass of blowfly maggots eating their way through rotting flesh is enough to make most people feel ill. Amazingly, this very technique has been used to make people well. Bad wounds often contain dead and decaying tissue that has to be cut out before the wound will heal. Doctors have found that maggots do the job much more efficiently than a surgeon's knife. Most maggots do not like healthy flesh, so when they run out of dead flesh they stop feeding. The result is a clean wound, with all the decay and infection stripped away.

I DIDN'T KNOW THAT

Glossary

abdomen rear section of an insect's body

antenna (more than one are called antennae) feelers on an insect's head used for touching and detecting tastes, smells, and sounds

bacteria microscopic organisms whose bodies are made up of a single cell. Some bacteria can cause illness in people and animals.

burrow to dig a hole or tunnel. A hole or tunnel used by an animal as a home is also called a burrow.

cell basic unit of structure and function in an organism

chitin hard substance that forms a fly's external skeleton and wings

chromosome in a cell, one of many tiny structures containing the genes

compound eye eye made up of thousands of tiny lenses and found mainly in adult insects

decomposer animal that helps to break down dead plants and animals

ectothermic having a body that warms up or cools down along with the outside temperature; cold-blooded

endothermic able to keep warm by producing heat inside the body using energy from food; warm-blooded

gene tiny packet of chemicals located on the chromosomes in a cell, which controls an aspect of an organism's appearance or behavior

habitat specific place where an organism lives

haltere one of two knobby growths on either side of a fly's body that helps it keep its balance during flight

host living thing that a parasite lives on or inside of

immune able to resist a disease

larva (more than one are called larvae) second, wormlike stage in the life cycle of most insects, after the egg and before the pupa. Also called a maggot or a grub.

mammal type of animal that is endothermic (warm blooded), has hair or fur, and nourishes its young on mother's milk

metamorphosis complete change in form that some animals (such as flies, frogs, and butterflies) go through to become adults

microbe organism too small to be seen without a microscope

mimic to imitate; an animal that imitates another animal is called a mimic

parasite living thing that lives on or inside of another living thing, taking nourishment from it and giving nothing in return.

parasitic living as a parasite

pendulum swinging weight that moves back and forth, as in an old-fashioned clock

pollen dustlike grains produced by a flower for reproduction purposes

pollinate to carry pollen from the male part of a flower to the female part, allowing it to produce a seed. This process is called pollination.

prey animal that is hunted and eaten by other animals for food

pupa (more than one are called pupae) third stage in the life cycle of most insects in which the young insect develops into the adult form inside a hard protective case. To pupate means to turn into a pupa.

species group of organisms that share certain features and that can breed together to produce offspring that can also breed

spiracle opening that allows air into the trachea of an insect

spore tiny, seedlike structure that can grow into a new living thing

tergum thickened plate on the back of an insect

thorax middle part of an insect's body containing muscles that move the wings and legs

trachea (more than one are called tracheae) tiny tube that sends air throughout an insect's body

ultraviolet kind of light that is invisible to humans

Further Reading

Berger, Melvin, and Gilda Berger. *Flies Taste with Their Feet: Weird Facts About Insects.* Illustrated by Robert Roper. New York: Scholastic, 1997.

Brimner, Larry Dane. *Flies.* Danbury, Conn.: Scholastic Library, 2000.

McEvey, Shane. *Flies.* Broomall, Penn.: Chelsea House, 2001.

Merrick, Patrick. *Biting Flies.* Eden Prairie, Minn.: Child's World, 2000.

Pascoe, Elaine. *Flies.* Photographs by Dwight Kuhn. Farmington Hills, Mich.: Gale Group, 2001.

Swan Miller, Sara. *Flies: From Flower Flies to Mosquitoes.* Danbury, Conn.: Scholastic Library, 1999.

Index

Numbers in *italic* indicate pictures.